I Know
Alike and Different

Reading consultant: Susan Nations, M.Ed.,
author, literacy coach, and consultant in literacy education

Printed in China

ISBN 13: 978-0-15-360212-2
ISBN 10: 0-15-360212-0

7 8 9 10 0940 16 15 14 13
4500409957

Harcourt
SCHOOL PUBLISHERS

These are alike.

These are different.

These are alike.

These are different.

These are alike.

These are different.

These are alike.

These are different.

These are alike.

These are different.

Glossary

alike

different

Photo credits: cover, pp. 2, 3, 4, 5, 6, 7, 8, 11, 12, 13 Gregg Anderson; p. 9, 10 © Hermera Technologies Inc.